Compendium of indicators for nutrition-sensitive agriculture

Food and Agriculture Organization of the United Nations
Rome, 2016

Contents

Foreword

2016 marks the beginning of the Decade of Action on Nutrition, which follows the Second International Conference on Nutrition (ICN2),[1] during which FAO Member Countries reaffirmed their commitment to end all forms of malnutrition.[2] These include a range of manifestations such as stunting, wasting, anaemia, and obesity. With the adoption by the United Nations General Assembly of the 2030 Agenda for Sustainable Development, a new indicator framework will guide the monitoring efforts of the international community in the period 2016-2030. The 2030 Agenda, and in particular the Sustainable Development Goal 2 (SDG2), recognize agriculture and food systems as major contributors to food security and nutrition. The experience of monitoring the Millennium Development Goals (MDGs) has shown that *what gets measured gets done* and that the *effective use of data can help to galvanize development efforts, implement successful targeted interventions, track performance and improve accountability.*[3] The SDGs, ICN2, and the Decade of Action on Nutrition call stakeholders – including Governments, donors, businesses, and civil society organizations – to take action and to track, report and evaluate their results (and investment) on efforts to improve nutrition across multiple sectors. In preparing these new policy frameworks, extensive discussions at global and country level have centred on agriculture and food systems, given that inadequate access to and consumption of healthy diets are common to all forms of malnutrition. This *Compendium of Indicators for Nutrition-Sensitive Agriculture* is grounded in the concrete needs of programme and project officers for harmonized and reliable monitoring instruments, by providing an overview of relevant indicators, along with recommendations on how to select the most appropriate ones, according to the economic and social context. Its purpose is to provide methodological information on currently available indicators that may be relevant for the Monitoring and Evaluation (M&E) of nutrition-sensitive agriculture investments. This document is the result of a fruitful collaboration between the Nutrition and Food Systems Division, the Investment Centre and the Statistics Division of FAO. It aims to complement other guidelines for nutrition-sensitive programme formulation, including the *Key Recommendations for Improving Nutrition through Agriculture and Food Systems*[4] and *Designing Nutrition-Sensitive Agriculture Investments: Checklist and Guidance for Programme Formulation.*[5] It is the result of a thorough process of review and extensive consultation within FAO and with development partners.

Anna Lartey
Director
Nutrition and Food Systems Division
FAO

Gustavo Merino
Director
Investment Centre
FAO

Pietro Gennari
Director
Statistics Division
FAO

Acknowledgements

This *Compendium* was written by Anna Herforth (FAO), Giorgia F. Nicolò (FAO), Benoist Veillerette (FAO) and Charlotte Dufour (FAO). Special thanks goes to Anna-Lisa Noack (FAO) and Sophia Lyamouri (FAO) for their valuable contributions throughout the entire process.
We are grateful to the following individuals who reviewed earlier drafts and generously provided feedback and inputs which improved the *Compendium*: Anna Lartey (FAO), Terri Ballard (FAO), Mary Arimond (UC Davis), Carlo Cafiero (FAO), Catherine Leclercq (FAO), Warren Lee (FAO), Catherine Bessy (FAO), Mary Kenny (FAO), Markus Lipp (FAO), Vittorio Fattori (FAO), Valentina Franchi (FAO), Marya Hillesland (FAO), Flavia Grassi (FAO), Pushpa Acharya (WFP), Emmanuelle Beguin (DFID), Heather Danton (USAID SPRING), James Garrett (IFAD), David Howlett (DFID), Aira Htenas (WB), Victoria Quinn (HKI), Rolf Klemm (HKI), Frederick Grant (HKI), Suneetha Kadiyala (LSHTM), Edye Kuyper (UC Davis), Yves Martin-Prével (IRD), William Masters (Tufts University), Quinn Marshall (WFP), Zalynn Peishi (DFAT), Victor Pinga (USAID SPRING), Melissa Williams (WB), Sara Marjani Zadeh (FAO), Camelia Bucatariu (FAO), Bin Liu (FAO). Finally, the support of Illia Rosenthal, editor (FAO), Juan Luis Salazar, graphic artist (FAO) and Chiara Deligia, communication officer (FAO) are warmly acknowledged.

The development of this *Compendium* was made possible with the support from the European Union through the Improved Global Governance for Hunger Reduction Programme.

1. Overview & purpose

In November 2014, during the Second International Conference on Nutrition (ICN2), the Member Countries of the UN Food and Agriculture Organization (FAO) and World Health Organization (WHO) adopted the Rome Declaration on Nutrition and its Framework for Action. By doing so, they committed to addressing all forms of malnutrition, including chronic and acute undernutrition, overweight and diet-related non-communicable diseases, and micronutrient deficiencies. Achieving these commitments requires reviewing the way food systems – the processes whereby food is produced, processed, transported, marketed and consumed – are being transformed. The ICN2 Framework for Action therefore places a strong emphasis on making food systems related policies and investments nutrition-sensitive. Governments and development partners are therefore increasingly including measures to ensure that food and agriculture investments and policies contribute to improved nutrition. These investments and policies cover a wide range of intervention areas, including value chain development, increase the food production, productivity and diversity, social and rural development.

An investment policy, programme or project can be considered nutrition-sensitive if it aims to contribute to better nutrition by addressing some of the underlying determinants of nutrition – access to safe and nutritious foods (quantity and quality/diversity), adequate care, and a healthy and hygienic environment. Such projects need to demonstrate that they lead to results toward improved nutrition.

This compendium has been designed to support officers responsible for designing nutrition-sensitive food and agriculture investments, in **selecting appropriate indicators** to monitor if these investments are having an impact on nutrition (positive or negative) and if so, through which pathways. It provides an overview of indicators that can be relevant as part of a nutrition-sensitive approach, together with guidance to inform the selection of indicators.

- **The purpose of this compendium** is to provide a current compilation of indicators that may be measured for identified outcomes of nutrition-sensitive investments. This compendium does not provide detailed guidance on how to collect a given indicator but points to relevant guidance materials.

- **This compendium does not represent official FAO recommendations for specific indicators or methodologies**. It is intended only to provide information on the indicators, methodologies and constructs that may be relevant to consider in the monitoring and evaluation of nutrition-sensitive agriculture investments.[a]

- It is not envisaged that a single project should collect data on all the indicators presented here. The selection will be informed by the type of intervention implemented, the anticipated intermediary outcomes and nutritional outcomes, as well as the feasibility of data collection in view of available resources and other constraints.

---

[a] In some cases, there is a construct for which there is no standard indicator or methodology. For example, income is an important construct for which there is not one precise gold standard methodology.

- The advice of M&E experts and subject matter specialists,[b] should be sought in making the final choice of indicators and in planning the data collection and analysis, including sampling and design of questionnaires.

- This compendium deals with programmes, projects and investments. While some indicators may be relevant for routine monitoring at national scale, this document does not cover every indicator that would be needed to monitor nutrition sensitivity of policies.

The document is composed of three parts:

1. The first part (chapters 2 to 5) introduces basic indicator categories, how they may be affected by common types of interventions and how the most appropriate indicators can be selected and integrated within a given project. It includes:

   a. A framework (Figure 1) by which indicators are organized, which identifies six outcome areas that are directly affected by agriculture, and how these can influence food access, diets and nutrition (in the next two sections, available indicators are compiled for each of these outcome areas).
   b. A matrix (Figure 2) of common investment/intervention types (agriculture, value chain, social development, irrigation, natural resource management) and how these can contribute to better nutrition through improving the six outcome areas.
   c. Basic tips on a sound identification of impact pathways for specific projects, so that the most appropriate indicators can be chosen.
   d. Considerations on the practicality of the modes of data collection when planning the M&E to capture the most appropriate indicators for assessing nutritional aspects.

2. The second part (chapter 6) is a summary of key indicators for nutrition-sensitive agriculture and food systems: currently available indicators that are recommended to measure each outcome area potentially affected by agriculture investments and policies.

3. The third part (chapter 7) is a longer compendium of indicators, including a description of what each indicator measures, when it is relevant to use it, how it is collected and analysed, and technical resources available related to it.

---

[b] E.g. If the project M&E will include dietary assessment, seek advice of nutritionists who have expertise in dietary assessment.

2. Simplified impact pathways from agriculture to nutrition

Figure 1 shows six outcome areas that are directly affected by an intervention, and how these can affect nutrition:

Figure 1. Simplified impact pathway framework of investment projects. This framework identifies six outcome areas that are directly affected by agriculture, rural development and food systems, and how these can influence nutrition (see glossary of terms, page10).

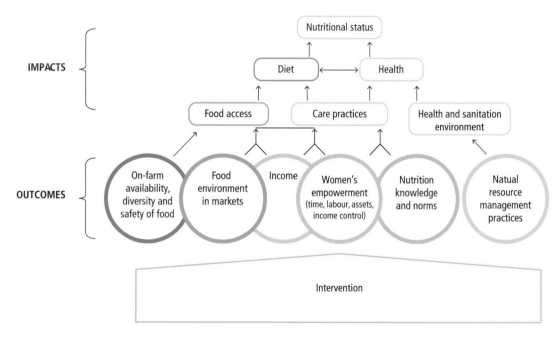

- **Food access** through improved access to nutritious foods on-farm, increased availability and lower prices of diverse nutritious foods in markets, improved food safety, and income which can be spent on more diverse nutritious food if such food is available, affordable and convenient.

- **Care practices** through empowerment of women (particularly if they can control income, their time and labor), and through incorporating behavior change communication.

- **Health and sanitation environments** through management practices that protect natural resources (water in particular) and safeguard against health risks introduced by agricultural production (e.g. livestock, standing water, agrochemicals).

Source: Herforth and Ballard, 2016.[6]

Note: see matrix of investment types (next pages) for examples of interventions.

3. Types of agricultural investments and entry points for nutrition

The above framework identifies six outcome areas that are directly affected by agriculture interventions or investments (the bottom row of "bubbles"); these affect the underlying determinants of nutrition (food access, care practices and health and sanitation environments), which affect diet and health and, ultimately, nutritional status.

Development organizations – such as FAO, the World Bank, the International Fund for Agriculture Development (IFAD), regional banks such as the European Bank for Reconstruction and Development (EBRD), the Global Environmental Fund (GEF) – and bilateral donors make several types of agriculture investments. These types of investments are represented in the matrix below (Figure 2), which shows how each is most likely to affect the six areas at the bottom row of Figure 1.

Some of the investments, if well-designed, can have impact pathways that are likely to contribute directly to some of the outcomes; they are highlighted in green. Others could affect those outcomes provided some nutrition-sensitive approach is applied; they are highlighted in yellow. Others do not typically affect those outcomes, unless some complementary, more nutrition specific intervention is added; they are in white.

The purpose of this matrix is to provide a few concrete ideas of how various investments may contribute to nutrition, leading to a clearer understanding on the types of intervention that may be implemented and which outcomes are most appropriate to measure. These are illustrative examples of entry points for these investment types. The diversity of country situations and projects makes it impossible to anticipate all possible entry points and contributions.

4

Figure 2. **Matrix of investment types and entry points for nutrition**

FOOD ACCESS, DIETS and health

Investment project types	Entry points	On-farm food availability & diversity	Food environment in markets	Income	Women's empowerment	Nutrition knowledge & norms	Health & sanitation environment
Agriculture development (extension research, area development inputs)	**Agriculture intensification** **Agriculture diversification** **Livestock and fisheries** **Extension** -Farmer field schools	Meet dietary gaps through own production	Increase availability and affordability of nutritious foods and diets in markets	Increase equitable access to resources and income; reduce poverty	Increase women's access to resources, know-how and income; reduce labour and time burden	Increase awareness/ Behaviour Change Communication (BCC) of nutritious foods and diets	Improve food safety, e.g. reduce mycotoxins & contamination (e.g. from agrochemicals)
Value chain development (including agro-processing)	**Storage & transportation** **Processing** **Trade & market linkages** **Marketing & promotion** -Nutrition focused marketing	Increase on-farm and off-seasonal availability of targeted nutritious crops	Increase variety in local markets, reduce prices & postharvest losses & improve convenience of nutritous foods	Increase income from value addition and technial expertise; reduce poverty	Increase women's access to resources, know-how and income; reduce labour and time burden	Increase awareness/ BCC of nutritious foods and diets and retaining nutrient content	Improve food safety, and food standards
Community-Driven Development (CDD)/Social development	**Rural institutional development** - Women's self-help groups - Capacity development **Social activities** - Community facilities - Social development/WASH **Financial inclusion/livelihood activities** - Income generating activities	Increase crop productivity and diversity food subsidies & distribution; households gardens	Strengthen storage, processing and retail of nutritious foods in markets	Increase equitable access to resources and income & enable savings and strategic investments; reduce poverty	Enable equitable decision-making; increase women's access to resources, know-how and income; reduce labour and time burden	Increase nutrition knowledge/BCC including awareness of healthy diets	Improve hygiene and sanitation practices and infrastructure
Water, irrigation and drainage	**Irrigation and drainage** **Water for domestic use** - Drinking water - Hygiene and sanitation **Water management**	Increase crop productivity and diversity and off-season production	Increase off-season availability & affordability of nutritious foods in markets	Increase crop production and income; reduce poverty	Reduce time burden from obtaining water		Reduce risk of waterborne and vector-borne disease; increase access to clean water
Natural resource management/ Forestry/ Environmental	**Biodiversity promotion** **Climate smart & nutrition sensitivity win-win** **Soil rehabilitation**	Sustain biodiversity for diet diversity; traditional indigenous and underutlized food species; Non-Timber Forest Products (NTFPs)	Increase availability of nutritious and underutilized foods in markets	Decrease risk of disasters/ catastrophic income loss (resilience)	Increase access to resources and income; reduce labour time and burden		Reduce environmental risks for food items (contamination)
Key		Green = important entry points to leverage and measure		Yellow = potential contribution requiring attention; measure if addressed		Blank = typically less of a direct contribution, although linkages may be possible; can be measured to ensure no harm	

4. Which indicators to choose: identifying impact pathways

The above matrix (Figure 2) shows how common investment types are most likely to affect the six outcome areas leading to improved food access, diets and health; and thus to show where impact should be estimated ex ante (through the financial and economic analysis), as well as during implementation through M&E. Important points to be considered are:

(1) an intervention would not typically be expected to affect all the outcomes depicted, and
(2) there is no automatic mechanism by which agriculture projects positively impact on nutrition, but there are plenty of potential entry points if one carefully designs these projects in a nutrition-sensitive manner. These possible entry points for nutrition-sensitive approaches are represented in the matrix above.

As the yellow and green colours in the matrix (Figure 2) suggest, some interventions are better suited to address certain impact pathways over others. For example, a project strengthening the value chain of a specific nutritious food (e.g. ground nuts) may have an impact on increasing the availability of

that food on farms and in markets, which could possibly lead to improved access to nutritious diets (food security) and improved diets. An irrigation project may have an entirely different impact pathway, through incorporating improved water sources for household use, thereby improving the health and sanitation environment, leading to reducing water-borne disease.

Each project or investment needs to be analysed *ex ante* for a clear theory of change and depending on the nature of the agriculture-nutrition intervention, the most appropriate type of indicators will vary. The matrix of investment types above illustrates where it may be most appropriate to use certain types of indicators. Indicators can be selected for each of the outcome areas depicted in Figure 1, based on expected impact pathways toward improved nutrition. Indicators for each outcome are compiled in the tables that follow. **To choose indicators, identify *which* of the outcomes the project is likely to affect, and *how* it will lead to improved food access, diet, and/or nutrition – in other words, the impact pathway and project results chain (see Figure 3 below).**

Figure 3. **Project results chain**

| **Project inputs** (staff hired, seeds obtained) | → | **Project outputs** (delivery of goods and services, e.g. trainings held, farmers reached) | → | **Intermediate outcomes** (e.g. increased production of ground nuts) | → | **Outcomes** (the 6 areas in the bottom row of Figure 1 and the top row of Figure 2) | → | **Impacts** (on food security, diet, and/or nutritional status) |

To affect these outcomes and impacts, the project needs to be implemented efficiently. **Process monitoring of the project inputs, outputs and outcomes can help improve plausibility that project activities are linked to results.** Process monitoring includes basic questions such as: *were inputs delivered and how? Who received them?* The scope of this document does not include presenting necessary indicators for process monitoring of programmes or projects, given the wide diversity of investments. Specific process indicators should be tailored to the activities of each programme or project.

Furthermore, nutrition sensitivity is also about ensuring that nutrition is not harmed. Ideally, an agriculture and rural development investment would aim to contribute to better nutrition. However, this is not always the case. Often an investment operation has been identified by a government (sometimes in agreement with an International Financing Institution, or IFI) with a particular objective such as agricultural productivity, value chain development, poverty alleviation, increased rural incomes or irrigation rehabilitation. These projects can be checked to ensure that they can result in improvement of – or at least not deterioration of – the underlying determinants of nutrition, by measuring the indicators most relevant to their activities.

This analysis of which indicators are most appropriate will be unique for each project or investment. However there are several general considerations for choosing indicators:

- many food and agriculture investments will affect production and/or consumption of nutritious food. **Indicators of food environment, food access and dietary quality** are often the most appropriate types of nutrition-relevant indicators for which improvements can be attributed to the investment intervention. Measures of the food environment include prices of nutritious foods in the market – something that many investments may affect but that often goes unmeasured.

- Many interventions will affect one or more aspects of **women's empowerment**, whether by design or not. Aspects including women's income control and time/labour burden should be assessed quantitatively or qualitatively, to ensure that the intervention does not cause harm to women themselves and does not place additional constraints on their choices about child care practices.

- Often, programmes are designed to improve **income** generation that can contribute to better nutrition. However, past research has shown that increased income alone does not automatically translate into better diets and nutrition. As indicated in Figure 1, the impact of income on diets depends on the food environment (what kinds of food are available, affordable, convenient and desirable), and also on who controls the income. The implication is that it may be useful to measure whether income has increased at household level, but it is also important to understand whose income has increased, and how this income is spent.

- Some interventions may affect natural resource management that affects people's exposure to health risks (i.e. the **health and sanitation environment**). For example, irrigation or livestock projects may affect drinking water quality. These are appropriate areas to measure in some projects.

- Many decision makers, programme managers and development partners wish to see impact on **nutritional status** indicators, such as stunting. However, it is difficult to observe and attribute impact on nutritional status from a particular investment intervention for two main reasons:

  - the targeted impacts of an agricultural intervention may not necessarily address the most significant causes of poor nutritional status in a given location. Depending on the context, other factors, such as low birthweight, inadequate breastfeeding and frequent infections may have stronger impacts on child growth than the amount or quality of food. Food access and diet quality are important in all contexts but might not be reflected in sudden changes in body size. Likewise, changes in women's empowerment, water quality, or other factors affected by agriculture may be important, yet may not be immediately reflected in anthropometry.

- Inadequate statistical power. Sample sizes required to observe a 5-10 percent reduction in rates of stunting, for example, are typically in the thousands or tens of thousands. The sample size required for adequate statistical power is often greater than the entire coverage of an intervention.

- For **practical reasons, it may be useful to prioritize indicators that are already collected in country, or that can easily be integrated** in existing surveys and data systems. In many cases, the most appropriate indicators are not already collected. Where there is limited capacity to collect data on key appropriate indicators, then it is important to plan resources for capacity development on the collection, analysis and use of these new indicators.

- When reducing **seasonal variability** of food access is a goal, data will probably need to be collected at several time points throughout the year. New indicators are generally not required; rather the same indicators can be collected at multiple time points.

5. Planning M&E to capture selected indicators

Most nutrition-relevant indicators (in chapters 6 and 7) would be collected through a household survey (often with a need for an individual respondent) that would need to be planned as part of the M&E system. This entails making financial provisions for the data collection and analysis, including the mobilization of necessary technical assistance to ensure that the data collected are of high quality.

- Useful considerations for preparing and budgeting for a household survey can be found in IFAD's *Results and Impact Management System (RIMS) handbook*.[7]

- Survey planning, coordination and implementation material as well as information on household surveys at country level are available at the International Household Survey Network.[8]

When household surveys are not part of the M&E plan, it may be challenging to collect information on how the project is affecting nutrition. Some nutrition-relevant information can be collected at community level or surveys of actors along the value chain or market surveys, including information on:

- the food environment (e.g. prices of nutrient-rich foods in markets; community-level production diversity);

- the health and sanitation environment (e.g. risk of water-borne disease in the community; quality of water supplied to communities, which could be affected by agriculture).

Glossary of terms in the framework

Nutritional status: in measurable terms, measures of anthropometry including child stunting (low height for age), wasting (low weight for height), underweight (low weight for age), body mass index (a proxy of body fatness), maternal underweight (low body mass index), or micronutrient status (measured by biochemical indicators).

Diet: the kind of food and drink an individual usually consumes.

Health: according to the WHO, health is a complete state of physical, mental and social well-being and not merely the absence of disease or infirmity. Notwithstanding this holistic definition, health is often operationally measured as the absence of communicable or non-communicable disease.

Food access: when people have physical and economic access to sufficient, safe, nutritious food to meet dietary needs, based on environmental and individual factors.

Care practices: an individual caretaker's typical practices for feeding and caring for infants, young children, mothers/selves and others in the family.

Health and sanitation environment: the factors in the physical environment where a person lives that pose health risks or protections.

Natural resource management practices: in measurable terms within this framework, practices undertaken to manage water, crops, fields and soils, biodiversity, or animals that may pose health risks or protections to households or individuals.

Food environment: the range of foods available, affordable, convenient and desirable to people.[9] Food market environments constrain and signal consumers what to purchase; wild and cultivated food environments also can provide access to foods. Fundamental elements of the food enviromnent are:

- *Availability: whether a food is present within a given geographic range.*
- *Affordability: price of a food, relative to cost of other foods and/or population income.*
- *Convenience: time and labour cost of obtaining, preparing, and consuming a food.*

- *Desirability: the external influences on how desirable a food is to consumers, including freshness/integrity of a food, how it is presented, and how it is marketed. This definition does not include intrinsic tastes/preferences of an individual, which influence consumption but are individual rather than environmental factors.*

On-farm availability, diversity and safety of food: the key elements of the on-farm food environment that affect people's access to diverse, nutritious, safe foods.

Income: cash and non-cash remuneration for work or investments, and gifts, received by a household or individual.

Women's empowerment: women's own power of self-determination and decision-making, including aspects of their control of assets, income, time, labour and knowledge.

Nutrition knowledge and norms: distinct from practices, knowledge that an individual has and social norms that affect caregiving and feeding/eating practices.

Intervention: project, programme or investment in agriculture, rural development and food systems. See Figure 2 for examples of intervention types.

6. Summary of key nutrition-sensitive indicators

These tables summarize key indicators that can be used to measure the outcome categories identified above. They are followed by a longer table, which compiles these key indicators plus many more, with a detailed description of where to find the methodology for that indicator (if a methodology exists) and for what it has been validated.

- Table 6.1 shows the top two recommended indicators at this point: Minimum Dietary Diversity for Women of reproductive age (MDD-W) as a measure of diet quality and Food Insecurity Experience Scale (FIES) as a measure of food access (it is an SDG2 indicator).

- Box 1 describes MDD-W in more detail, including how it is relevant to measure in agriculture projects and what its limitations are.

- Table 6.2 highlights indicator constructs that are important to measure, but for which a well-defined indicator or standard methodology may not necessarily exist. For example, there may be several methodologies to measure household income – a key indicator – but no precisely-defined gold standard method.

- Note that this table does not include indicators of care practices and nutritional status. Many agriculture investments do not directly impact care practices and nutritional status, although some individual projects may indeed aim to affect them and therefore measure them. Further information on indicators for these categories is available in the detailed tables in chapter 7.

Table 6.1 **Recommended Indicators**

Type of measure	Indicators	What the indicator measures	Resources	Mode of collection
Diet – Individual level	**Minimum Dietary Diversity for Women of reproductive age (MDD-W)** Minimum Dietary Diversity for young children (MDD age 6-23 months)	A measure of dietary quality, which reflects overall nutrient adequacy and dietary diversity. It does not reflect adequacy of specific target nutrients.	*Minimum Dietary Diversity for Women: A Guide to measurement.* (FAO/Family Health International (FHI) 360, 2016).[10] Indicators for assessing infant and young child feeding practices (WHO, 2008[11] and WHO, 2010[12]).	Household survey (individual interview within household)
Food access – Household level	**Food Insecurity Experience Scale (FIES)**	Severity of food insecurity experience within a household. Can also be measured for individuals.	Description of the indicator available at the Voices of the Hungry website.[13]	Household or individual survey

Table 6.2 Measurable outcomes for which various methods are available

Type of measure	Indicators	Resources	Mode of collection
On-farm availability, diversity, and safety of foods	**Production of target nutrient-rich foods**	There are various ways production of target nutrient-rich foods could be defined and measured, such as change in production volume, but no standard methodology.	Household survey or farm survey
	Diversity of crops and livestock produced	There is no standard method for measuring on-farm diversity for nutritional purposes. Three methods that have been used in the literature include: 1. simple count of species produced over the last 12 months (crops, plants and animals); 2. Shannon Index;[14] 3. Simpson Index.[15]	
	Months of Adequate Household Food Provisioning (MAHFP)	There is a MAHFP indicator guide available from Bilinsky and Swindale, 2010.[16]	
Food environment in markets	**Availability and prices of targeted nutrient-rich foods in local markets**	There are various methods for monitoring availability and prices of foods in markets but no standard methodology; see Table 7.4.	Market / Price information systems when they exist; or rapid market survey
Income	**Income, disaggregated by gender, to reflect intra-household income control**	There are various methods for constructing indicators to reflect household and individual incomes; see Table 7.5.	Household survey and/or enterprise records kept by project
Women's empowerment	**Women's access and control over resources** (e.g. land/property ownership)	There are various methods for constructing indicators to reflect these constructs, including time use surveys, qualitative inquiry, and some newer indexes; see Table 7.6 for detail.	Household survey and/or qualitative process
	Women's participation in economic activities (e.g. gender gap in crop/livestock sales)		
	Women's access to and control over benefits (e.g. agricultural income earned and controlled by women)		
Nutrition (and food safety) knowledge and norms	*(Indicators will be project-specific)*	There are guidelines available with questionnaires on knowledge, attitudes and practices related to most common nutrition topics (Fautsch Macías and Glasauer, 2014).[17]	Household survey and/or qualitative process
Natural resource management practices	**Access to improved drinking water source** (see Table 7.9 for indicator definitions)	The WHO/UNICEF Joint Monitoring Programme has established a standard set of drinking-water and sanitation categories that are used for monitoring purposes.[18]	Farm survey

Box 1. Minimum Dietary Diversity-Women (MDD-W): an indicator of dietary adequacy that is relevant for agriculture

The MDD-W responds to a long-standing need to have a simple and effective indicator to assess women's diet quality. Women are a group that is often nutritionally vulnerable because of their increased requirements in micronutrients and because, in some settings, they may be disadvantaged in intra-household distribution of nutrient-dense foods. Nutrition-sensitive programming in agriculture has intensified in recent years due to an increased focus on deploying efforts towards good nutrition for women and children during the critical 1,000-day period of their life. The MDD-W offers one way to measure impact of these nutrition-sensitive efforts.

The MDD-W is a brief set of questions, requiring much less time and expense than traditional dietary surveys, which can be included in M&E systems. It is validated as an indicator of nutrient adequacy. Moreover, it can provide information about dietary patterns and what are the food groups predominantly consumed at population level (or missing from the diet) and in a given agro-ecological zone. For example, indicators can be derived for consumption of vitamin A-rich plants, and for consumption of iron-rich food groups. This information, if properly accessed and incorporated to inform decision making, can provide sound evidence to influence policies and investment choices towards more nutrition-sensitive agriculture production.

It is important to note that MDD-W does not provide comprehensive information on diet quality or all impacts of agriculture on diet. It may not capture changes when projects aim to increase production and consumption of food items or food groups already widely consumed. Likewise, it will not reflect increase in nutrient intake due to consumption of fortified or biofortified foods. These projects can have a positive impact on nutrition but need other metrics. Also, it does not measure consumption of unhealthy foods such as ultra-processed snacks and sugar-sweetened beverages, which negatively affect diet quality and non-communicable disease risk in many settings.

MDD-W is a powerful tool to track progress and raise awareness on gender-specific needs and it fosters the message of the important link between food production (agriculture) and individual consumption (nutrition). As always, project managers need to be aware of the information this indicator does and does not reflect, and to choose indicators that are appropriate to reflect their project inputs and impact pathways. MDD-W is one useful, validated indicator that can measure progress toward improving diet quality.

7. Detailed compendium of existing indicators for nutrition-sensitive investments

This section provides a compendium of existing indicators for each area in the *Simplified impact pathway framework of investment projects* (Figure 1), including a description of what each indicator measures, when it is relevant to use it, how it is collected and analysed, and technical resources available related to it. The purpose of this compendium is to provide a current compilation of indicators that may be measured for identified outcomes of nutrition-sensitive investments. This compendium does not provide detailed guidance on how to collect a given indicator but points to relevant guidance materials.

More detailed information on the background and validation for several of these indicators can be found in the Food Security Information Network (FSIN) publication (Lele, U. *et al.* 2016).[19]

Table 7.1 Diet quality – Individual level

- **When to use:** if the intervention affects food environments or income, women's empowerment and/or nutrition knowledge, skills and practices with hypothesized impact on diet quality.
- **Note**: no easy indicator currently exists that can capture diet quality holistically in its entirety (i.e. a diet that follows dietary recommendations). The MDD-W is validated and relatively easy to administer, but it does not capture dietary quality completely because it is an indicator of micronutrient adequacy and diversity, but does not deal with unhealthy amounts or components of the diet. Other dietary quality scores have been constructed (e.g. the Healthy Eating Index, Dietary Quality Index), but these require a full quantitative 24-hr recall. More diet quality indicators are under development. Currently there are several indicators that capture some aspects of diet quality:

Indicator	What it measures	Population	Data collection	Data analysis	Notes
MDD-W (Minimum Dietary Diversity – women of reproductive age)	A measure of dietary quality, which reflects nutrient adequacy and dietary diversity	Women of reproductive age (15-49 years)	Data are collected on the foods and beverages consumed in the previous 24 hours which are aggregated into 10 distinct food groups. Does not require quantitative food intake.	Several indicators can be derived from the basic data, including (i) proportion of women who consume 5 or more food groups out of ten; (ii) mean dietary diversity score; (iii) proportion of women consuming any specific food group such as animal source foods.	**VALIDITY** This indicator has been validated as an indicator of likelihood of micronutrient adequacy among women of reproductive age. There is a recent global consensus on this indicator as the best, most valid measure of women's dietary diversity; it replaces the WDDS (Women's Dietary Diversity Score) that had been previously developed by FAO and Food And Nutrition Technical Assistance project (FANTA). Unlike former measurements, it offers a threshold for women's micronutrient needs. Consortium of International Agricultural Research Centers (CGIAR) and USAID Feed the Future have mainstreamed the use of this indicator in their evaluations. **CUTOFF** (available) Women who consume foods from at least 5 out of 10 food groups have a higher likelihood of micronutrient adequacy. **METHODOLOGY** (standardized) Standardized methodology for data collection and analysis is available from FAO/FHI 360, 2016.[10]

Indicator	What it measures	Population	Data collection	Data analysis	Notes
Minimum Dietary Diversity – Young children	A measure of dietary quality, which reflects nutrient adequacy and dietary diversity feeding practices	Children under 2 years	Same as above. The guidelines recommend open recall but Demographic and Health Surveys (DHS) use a list	Proportion of children 6-23 months of age who receive foods from 4 or more food groups (of 7) It is recommended that the indicator be further disaggregated and reported for the age groups: 6–11 months, 12–17 months and 18–23 months	**VALIDITY** Consumption of foods from at least 4 food groups out of 7 on the previous day would mean that, in most populations, the child had a high likelihood of consuming at least one animal-source food and at least one fruit or vegetable that day, in addition to a staple food (grain, root or tuber). **CUTOFF** (available) The cutoff of at least 4 of the above 7 food groups above was selected because it is associated with better quality diets for both breastfed and non-breastfed children. **METHODOLOGY** (standardized) This indicator is a result of discussions by a large technical stakeholder group – WHO, UN Chilidren's Fund (UNICEF), USAID, University of California (UC Davis), the International Food Policy Research Institute (IFPRI) – and has been published by WHO, 2008.[11]
Individual Dietary Diversity Score (IDDS)	A measure of dietary quality, which reflects nutrient adequacy and dietary diversity	Usually children over age 2 years	Consists of either an 8-question list (one for each food group), or a qualitative 24-hour food list (i.e. what was eaten by the child yesterday, without amounts)	Sum score – can calculate a mean or percentiles	**VALIDITY** This indicator has not been validated as a measure of micronutrient adequacy, and it has been defined by FANTA. It has been used for children age 2-14 years, which is an age range that lacks a validated indicator of dietary diversity. **CUTOFF** No cutoff is defined in this indicator. **METHODOLOGY** This indicator is found in Swindale and Bilinsky, 2006.[20]
Unique Food Items/ Dietary variety	A proxy for dietary quality	Individual	Qualitative 24-hour food list (i.e. what was consumed by the respondent yesterday, without amounts)	Count of the unique food items consumed	**VALIDITY** Validity of food variety as an indicator of nutrient adequacy has been assessed with a food frequency questionnaire for Western Mali (Torheim, *et al.* 2003).[21]

Indicator	What it measures	Population	Data collection	Data analysis	Notes
Quantitative nutrient intakes	This is the most detailed measure when the primary concern is specific information on precise nutrient intakes	Individual	Quantitative 24-hour recalls (i.e. what was consumed by the respondent yesterday, using methodology to acquire amounts consumed), weighed food records or diaries (estimated food records)		**VALIDITY** **Quantitative 24-hour recalls**: assess average usual intakes of a large population provided that the sample is truly representative and that the days of the week are adequately represented. Multiple replicate 24-hour recalls are needed to estimate usual intakes of individuals. Can be used with illiterate individuals. **Weighed food records**: access actual and usual intakes of individuals, depending on number of measurement days. Accurate, time consuming and expensive. Requires literate participants. **Estimated food records**: assess actual and usual intake of individuals. Accuracy depends on the conscientiousness and ability of subjects to estimate quantities. Requires literate participants. **CUTOFF** Nutrient intakes can be compared to recommended daily intakes in order to derive information on: (i) mean nutrient intake of a group; (ii) percentage of population "at risk" of inadequate nutrient intake; (iii) ranking individuals by food or nutrient intake. **METHODOLOGY** Methodological guidance for measuring food consumption of individuals can be found in: *Principles of Nutritional Assessment* (second edition). Gibson (2005).[22] A useful document on methodology for multi-pass 24-hour recalls, available from Gibson and Ferguson 2008.[23] Note: this indicator is more time-intensive than others and requires significant training of enumerators to collect data and time/funds for data analysis Note: the Agriculture for Nutrition and Health (A4NH) programme managed by the CGIAR uses the indicator *Dietary intake of selected micronutrients.*[24]

Indicator	What it measures	Population	Data collection	Data analysis	Notes
Consumption of 400g fruits and vegetables per day	Whether individuals are meeting the WHO recommendations for fruit and vegetable consumption	Individual	Quantitative 24-hour recalls, weighed food records or diaries (see methodologies above)	Sum the gram total of fruits and vegetables consumed in the previous day	**VALIDITY** Using the techniques for measuring quantitative food intakes, this would be a valid indicator of its definition: whether an individual consumes the recommended amount of fruits and vegetables. **METHODOLOGY** See above for measuring quantitative food intakes
Proportion of the diet consisting of processed/ultra-processed foods	Useful when chronic disease and obesity are concerns. A lower proportion may be associated with improved dietary quality related to risk of chronic disease (Monteiro et al., 2013)[26]	Individual	Quantitative food consumption surveys, either at household or individual level	This indicator has been constructed in terms of % calories from ultra-processed products	**VALIDITY** Methods are experimental at this point. **METHODOLOGY** Guidelines on the collection of information on food processing through food consumption surveys (FAO, 2015).[25] Note: this guide does not define an indicator. **DEFINITIONS** Monteiro et al. (2013)[26] define "ultra-processed" foods as "food products manufactured from industrial ingredients resulting from the extraction, refinement and modification of constituents of raw foods with little or no whole food." The International Agency for Research on Cancer (IARC)definition of "highly processed" foods: foods that have been industrially prepared, including those from bakeries and catering outlets, and which require no or minimal domestic preparation apart from heating and cooking (such as bread, breakfast cereals, cheese, commercial sauces, canned foods including jams, commercial cakes, biscuits and sauces). Moubarac et al. (2014)[27] define four categories of processing: i) unprocessed and minimally processed foods; ii) processed culinary ingredients; iii) processed foods; and iv) ultra-processed food and drink products.

Indicator	What it measures	Population	Data collection	Data analysis	Notes
Vitamin A-rich food consumption	Useful when vitamin A-rich foods are targeted and/or when vitamin A intake is of primary concern	Individual	At a household or individual level, requires a household survey	Many kinds of indicators could be used or created as appropriate to the specific intervention. Examples include (i) number of vitamin A-rich foods consumed at least once over a specified period; (ii) mean frequency of consumption of vitamin A-rich foods over a specified period	**METHODOLOGY** Depending on the indicator selected, data could be gathered using 24-hour qualitative recall methodology or a food frequency questionnaire. These are quicker alternatives compared to quantitative intakes from a quantitative 24-hour recall (see above *Quantitative nutrient intakes*). One food frequency method: the Helen Keller International (HKI) Food Frequency Method generates information about the availability, accessibility, preparation and seasonality of foods. It creates scores combining food groups of yellow/orange flesh fruits or vegetables, dark leafy green vegetables, in order to provide information on frequency of consumption of vitamin A-rich foods as well as information on feeding practices. It may underestimate vitamin A intake for young children consuming breastmilk and other milk. A tool is available online.[28] **DEFINITION** The Codex Alimentarius Guidelines[29,30] provide thresholds for considering a food as a "source" or a "high source" of different nutrients, based on the percent of the Nutrient Reference Value (NRV) provided by the food. A food must provide 15% of NRV per 100 grams to be considered a "source" of the nutrient. A food must provide double the "source" threshold, i.e. 30% of NRV per 100 grams, to be considered a "high source" of the nutrient.
Iron-rich food consumption	Useful when iron-rich foods are targeted and/or when iron intake is of primary concern	Individual	At household or individual level, requires a household survey	There could be many indicators to measure this concept. One is specifically designed for young children: "proportion of children 6–23 months of age who receive an iron-rich food or iron-fortified food that is specially designed for infants and young children, or that is fortified in the home."[11,12]	**METHODOLOGY** Depending on the indicator selected, data could be gathered using 24-hour qualitative recall methodology or a food frequency questionnaire. These are quicker alternatives compared to quantitative intakes from a quantitative 24-hour recall (see above "quantitative nutrient intakes"). **DEFINITION** The Codex Alimentarius Guidelines[29,30] provide thresholds for considering a food as a "source" or a "high source" of different nutrients, based on the percent of the Nutrient Reference Value (NRV) provided by the food. A food must provide 15% of NRV per 100 grams to be considered a "source" of the nutrient. A food must provide double the "source" threshold, i.e. 30% of NRV per 100 grams, to be considered a "high source" of the nutrient.*

Indicator	What it measures	Population	Data collection	Data analysis	Notes
Consumption of specific target foods	Useful to track whether individuals are consuming foods promoted by an intervention, or regardless of an intervention	Individual	At household or individual level, requires a household survey		**METHODOLOGY** Depending on the indicator selected, data could be gathered using 24-hr qualitative recall methodology or a food frequency questionnaire. **DEFINITIONS** Feed the Future (FTF) Indicator Handbook** has defined three examples of this kind of indicator.[28] Many others could be created that are appropriate to the specific scope and desired outcomes of an intervention. Examples include (i) if any of the specific food was consumed over a specified period (e.g. 1 day, 1 week); (ii) how frequently the specific food was consumed over a specified period (e.g. through a food frequency questionnaire); (iii) how much of the specific food was consumed over a specified period (quantitative intake in grams); (iv) diversity of consumption of foods within a food group over a specified time period (e.g. diversity of fruits and vegetables consumed).

* Defining "iron-rich" foods can be debatable. The Infant and Young Child Feeding (IYCF) indicators guide,[11,12] which was designed for infants under age two years, defines them as "flesh foods, commercially fortified foods specially designed for infants and young children that contain iron, or foods fortified in the home with a micronutrient powder containing iron or a lipid-based nutrient supplement containing iron." These foods have highly bioavailable iron, but the definition excludes plant sources of iron, which can also contribute to iron intake. This definition was not designed to be extrapolated to other age groups.

** FTF has recently developed nutrition-sensitive indicators to complement the dietary diversity indicators already being collected. The commodities included in these indicators must be nutrient-rich, i.e. meet any of the following criteria: i) bio-fortified; ii) legume, nut or seed; iii) animal-sourced food; iv) dark yellow or orange-fleshed root or tuber; v) fruit or vegetable that meets the threshold for being a "high source" of one or more micronutrients on a per 100 gram basis.

- **Prevalence of women of reproductive age who consume targeted nutrient-rich value chain commodities**. This is a population-based indicator of an outcome of nutrition-sensitive value chain interventions that measures the percent of women of reproductive age (15-49 years old) in United States Governement (USG)-assisted areas who consumed in the previous day one or more nutrient-rich commodities or products made from nutrient-rich commodities being promoted by USG-funded value chain activities. This indicator complements the Feed the Future indicator that captures increased dietary diversity among women of reproductive age.
- **Prevalence of children 6-23 months who consume targeted nutrient-rich value chain commodities.** This is a population-based indicator of an outcome of nutrition-sensitive agriculture interventions that measures the percent of children 6-23 months of age in USG-assisted areas (e.g. the Feed the Future Zone of Influence) who consumed in the previous day one or more nutrient-rich commodities or products made from nutrient-rich commodities being promoted by USG-funded value chain activities. This indicator complements the Feed the Future infant and young child feeding indicator.

Table 7.2 Food access – Household level

- **When to use:** if the intervention affects food production, income, seasonal variation of food access and prices.
- While there are many existing food security metrics, a suite of indicators that measures each dimension of food security (sufficiency, quality, acceptability, safety, certainty/stability) is not yet established (Coates, 2013).[39]

Indicator	What it measures	Population	Data collection	Data analysis	Notes
Food Insecurity Experience Scale (FIES)	Severity of food insecurity experience	Household or individual	8 question survey module	Thresholds set on the score to classify the severity status of respondents	**VALIDITY** The FIES has been collected in over 145 countries since 2014 in the Gallup World Poll. Each country dataset has been validated with the Rasch model (Item Response Theory), demonstrating that the scale is capturing the latent trait of food insecurity (access dimension). Statistical techniques have been developed to equate country results against a global standard that allows comparison across all countries. The global data reveal that the FIES shows significant and high correlations in the expected direction with most accepted indicators of development, including child mortality, stunting, poverty measures and the Gini index. **METHODOLOGY** (standardized) Description of the indicator available at the Voices of the Hungry website.[13]

Indicator	What it measures	Population	Data collection	Data analysis	Notes
Household Dietary Diversity Score (HDDS)	Household access to and consumption of a variety of foods	Household	Consists of a simple count of the different food groups that a household or an individual has consumed over the preceding 24 hours. Data are collected on the foods and beverages consumed in the previous 24 hours to ascertain if anyone in the household consumed any item from different food groups.	Foods consumed at household level aggregated into twelve food groups. Mean score.	**VALIDITY** The Household Dietary Diversity (HDD) indicator has not yet been tested for its performance in predicting micronutrient adequacy and should therefore not be used as an indicator of dietary quality at the household level, although it can be a useful indicator of food access. It excludes food eaten outside the home so information may be missed. **CUTOFF** (not available) There is no established number of food groups to indicate adequate or inadequate DD for the HDDS. However, for a project with interventions to improve food access and household food security, the mean HDDS of the wealthiest tertile could be used to set the HDDS target level. **METHODOLOGY** (standardized) *Guidelines for measuring household and individual dietary diversity* (FAO, 2012a).[32] Note that in this publication, the HDDS methodology was adapted from Swindale and Bilinsky (2006)[20]; and the WDDS described in it is now replaced by the new MDD-W indicator – see above. In short: use this publication for household dietary diversity. Use MDD-W for women's dietary diversity.[10]

Indicator	What it measures	Population	Data collection	Data analysis	Notes
Food Consumption Score (FCS)	Household access to consumption of diverse food; weighted by nutrient density	Household	Information about frequency of consumption (in days) by a household over a recall period of the past seven days is collected from a country-specific list of food groups	The score is calculated using the frequency of consumption of different food groups consumed by a household during the 7 days before the survey	**VALIDITY** The FCS has been validated against per capita calorie consumption within the household and several alternative indicators of household food security (percentage expenditures on food, asset and wealth indices). The food consumption score is being used widely by WFP in their surveillance activities. **THRESHOLDS** (available) The thresholds for the Food Consumption Groups (FCGs) should be determined based on the frequency of the scores and the knowledge of the consumption behaviour in that country/region. The typical thresholds are: 0-21 Poor; 21.5-35 Borderline; > 35 Acceptable. **METHODOLOGY** (standardized) Technical Guidance Sheet - Food Consumption Analysis: Calculation and Use of the Food Consumption Score in Food Security Analysis (WFP-VAM, 2008).[33]
Household Food Insecurity Access Scale (HFIAS)	Severity of food insecurity experience, requiring local adaptation	Household	9 questions in 4 domains, survey module	Responses may be categorized into 4 levels, or summed into a score ranging from 0-27	**VALIDITY** This indicator must be adapted to the local situation. It may not be valid without adaptation. **METHODOLOGY** Coates, Swindale and Bilinsky, 2007[34] available online.
Escala Latino-americana y Caribeña de Seguridad Alimentaria (ELCSA)	Severity of food insecurity experience, cross-culturally valid in Latin America and the Caribbean	Household	15 questions in 4 domains, survey module (8 questions refer to adults, 7 refer to children)	Responses may be categorized into 4 levels, or summed into a score ranging from 0 to 15	**VALIDITY** The ELCSA was developed taking into account previously validated food insecurity assessment scales at household level (US Household Food Security Supplement Module, *Escala Brasileña de Inseguridad Alimentaria* (EBIA), among others). **CUTOFF** (available) Different cutoff points referring to the level of food insecurity. **METHODOLOGY** (standardized) Manual by FAO (2012b)[35] available online.

Indicator	What it measures	Population	Data collection	Data analysis	Notes
Household Hunger Scale (HHS)	Cross-culturally valid measure of the severity of food insecurity experience	Household	3-question survey module	Thresholds set on the score (ranging from 0-6) to classify the severity status of respondents	**VALIDITY** This indicator is a cross-culturally valid indicator of hunger and has demonstrated the potential for both internal and external validity, with strong relationship with household income and wealth scores. It is most sensitive to severe food insecurity (hunger), and is less useful in situations of moderate or mild food insecurity. **CUTOFF** (available) Different cutoff points refer to the level of food insecurity. **METHODOLOGY** (standardized) *Household Hunger Scale: Indicator Definition and Measurement Guide.* Ballard *et al.* (2011).[36] **NOTES ON USE** The HHS is being used by USAID Feed the Future projects.
Coping Strategies Index (CSI)	Severity of food insecurity experience, requiring local adaptation. Used to identify vulnerable households and estimate long-term changes in food security	Household	A locally-adapted list of coping strategies and their severity weightings, is obtained through focus group discussions		**VALIDITY** Not clearly demonstrated across contexts, but useful for understanding how people respond to lack of food. **METHODOLOGY** (standardized) Manual by WFP-VAM (2008)[37] available online. **NOTES ON USE** The CSI has been used by the World Food Programme (WFP), CARE International and other NGOs.

Indicator	What it measures	Population	Data collection	Data analysis	Notes
Months of Adequate Household Food Provisioning (MAHFP)	Measures perceived household food adequacy throughout the past year and reflects the seasonality aspect of food security	Household		Sum total of number of months in the past year the household had inadequate food	**VALIDITY** Not clearly demonstrated across contexts, but useful for understanding seasonality of food security. **CUTOFF** No cutoff is available but targets could be established based on the months of adequate food provisioning of the top tercile (one-third) of the households or the average months of adequate food provisioning of the richest income tercile. **METHODOLOGY** (standardized) Available from Bilinsky and Swindale (2010)[16] **NOTES ON USE** It has been incorporated as a standard impact indicator in all Africare's food security programs.

Additional resources on the above indicators have been published by Jones *et al.,* 2013[38], Coates, 2013[39], FAO and WFP, 2012[40].

Table 7.3 On-farm availability, diversity and safety of food

- **When to use**: if the intervention affects the amount, type, or quality of food produced for home consumption.

Indicator	What it measures	Population	Data collection	Data analysis	Notes
Availability of specific foods on-farm	Useful to track whether specific foods of interest are available, such as those promoted by an intervention	Household or community	Household survey or observation		There are various ways this indicator could be defined, such as "availability of micronutrient-rich target foods on farms: Increased / decreased production in volume, across seasons and % compared to without project". USAID uses the indicator "total quantity of targeted nutrient-rich value chain commodities set aside for home consumption by direct beneficiary producer households," found in FTF, 2016.[31]
Diversity of foods produced on-farm	A measure of availability of diverse nutritious foods	Household or community	Household survey or observation		**METHODOLOGY** There is no standard or validated method for measuring on-farm diversity for nutritional purposes. Three methods that have been used in the literature include: 1. simple count of species produced over the last 12 months (crops, plants and animals); 2. Shannon Index;[14] 3. Simpson Index.[15]
Functional diversity index	A measure of availability of diverse nutritious foods	Household or community	Household survey and observation	See Remans *et al.* 2011 →	**METHODOLOGY** Assessing Nutritional Diversity of Cropping Systems in African Villages. Remans *et al.* (2011).[41]
Proportion of staple crop production that is biofortified	A proxy for micronutrient density of staple crops produced on farm	Household or community	Household or community survey		This is not a standard validated measure but could be used in projects that seek to increase micronutrient intake via biofortified crop production.

Indicator	What it measures	Population	Data collection	Data analysis	Notes
Implementation of good agricultural practices*	Safety of agricultural production targeted by project**	Household or community	Farmer surveys or observation to capture KAPs (Knowledge Attitudes and Practices)		Indicators will be project specific. Specific practices that improve safety of food production will depend on the nature of the production systems. These practices could be related for example to pesticide or veterinary drugs use; value chain specific cultivation practices; storage practices on farm; other hygiene practices (washing of agricultural products). If there is set of legislated standard practices, an example of indicator could be: % of compliance of primary producers to practices; increase of % of primary producers certified.
Grain loss***	Post-harvest loss	Community, farm and field levels			No uniform concepts, definitions and measurement techniques have been used in different studies estimating losses. This review of methods available for estimating grain loss covers techniques to estimate losses during harvesting, stacking, threshing/shelling, cleaning, drying, storage, transport, processing, packaging and/or due to insects, mould and pests. More information is available from the Global Strategy to improve Agricultural and Rural Statistics (GSARS).[42]

*Good Agricultural Practices (GAPs) are an essential prerequisite to improve food and feed safety from farm to plate and their application can be measured as a proxy indicator for safe food and feed production. However, it is important to keep in mind that, adoption of GAPs alone is not a guarantee that products are free from contaminants as process standards might or might not influence the characteristics of the end products. Expensive analytical techniques are the only methods to detect the presence of contaminants and therefore guarantee safety of food and feed. It is therefore advisable that, when identifying food safety indicators within programmes and projects, a risk-based approach is used which considers the full production process, from farm to plate of food and feed. In this regard, it is recommended to contact a local food safety expert, to help project officers apply this approach from the early stages of the design phase of programmes and projects.

**Chemical contaminants can be present in food and feed as a result of the use of agrochemicals, such as residues of pesticides and veterinary drugs, contamination from environmental sources (water, air or soil pollution), cross-contamination or formation during food processing and natural toxins.

*** Some of the methods developed for grains could potentially be applied to other crops, and/or new crop-specific methods could be developed.

Table 7.4 Food environment in markets

- **When to use**: if the intervention affects food availability, prices, marketing or safety; or to understand how income is likely to translate to food purchases.
- Indicators capturing availability, affordability, convenience and desirability of diverse foods in markets are currently few.

Indicator	What it measures	Population	Data collection	Data analysis	Notes
Availability of specific foods in markets	Useful to track whether specific foods of interest are available, such as those promoted by an intervention	Market	Market / Price information systems when they exist; rapid market survey if not, at a point in time or over seasons/ Surveys of ac-tors along the value chain		**METHODOLOGY** (not standardized) There are various ways this indicator could be defined, such as "availability in markets of foods promoted by investment (volume/across seasons)". Note: depending on the intervention activities, it may be appropriate to add indicators relevant to agricultural processes to increase availability of nutrient-rich foods: e.g. Reduced post-harvest losses of nutrient-rich foods; implementation of processing techniques that retain nutritional value.
Prices of specific foods in markets	Useful to track whether specific foods of interest are affordable, such as those promoted by an intervention.	Market	Market / Price information systems when they exist; rapid market survey if not, at a point in time or over seasons.		There are various ways this indicator could be defined, such as "prices of foods promoted by investment in project areas compared to areas without project".
Food prices	Useful to track whether a basket of foods is affordable.	Market	Market / Price information systems when they exist; rapid market survey if not, at a point in time or over seasons.		Often the price of a basic food basket is tracked, typically not based on nutritious diets. Prices of staple grains are often monitored by FAO[43] and WFP-VAM.[44]
Cost of a healthy diet	The minimum cost of a diet meeting minimum requirements of macro and micronutrients or food-based dietary guidelines.	Community	Methodology is not standardized. Sample methodology is published by Save the Children (see notes).	Linear programming	Save the Children piloted an approach "to quantify the extent to which households could afford to feed their children under the age of 2 and a whole family of 5 people, with a diet meeting minimum requirements of macro and micronutrients."[45] Additional resources are published by USAID.[46]

Indicator	What it measures	Population	Data collection	Data analysis	Notes
Functional diversity index	A measure of access to diverse nutritious foods.	Household or community		See Remans *et al.* 2011 ⟶	Indicator description can be found in Remans *et al.* (2011).[41]
Indicators of food safety within the food environment*		Market	Sample collection at market level		Specific indicators are not well defined, but could include: - % reduction of chemical or microbiological contaminants in products offered to consumers at retail - % compliance of product with national regulations for a specific product Sampling guidance tools are available online.[47,48] Note: Representative samples might be very costly and important variations might occur between places/timing for sampling.
Food loss in the supply chain	The amount of decrease in safe and nutritious food mass available for human consumption in the different segments of a specific supply chain.	Supply chain	Survey of producers, processors or handlers/sellers and other knowledgeable persons of the supply chain being assessed, complemented with ample and accurate observations and measurements and a literature review.	Results include qualitative and quantitative elements.	Specific indicators are not well defined, but some techniques are available for estimating food loss along the supply chain: Global Initiative on Food Loss and Waste Reduction (SAVE FOOD) field case study methodology.[49]

* Implementation of good hygiene practices in food production can play a large role in food safety. Specific indicators are not well defined and would depend on project context and interventions. The methodology could include surveys of actors along the value chain. Additional resources can be found online:

• FAO food safety and quality website.[50]

• Recommended International Code Of Practice - General Principles Of Food Hygiene.[51]

• Codex Alimentarius standards, guidelines and advisory texts.[52]

Table 7.5 Income

- **When to use**: if the intervention affects household income, which is hypothesized to affect food or health care purchases.
- Methodology depends on context: in some places, people can report household income; in other places, own production accounts for a substantial proportion of income, so it must be imputed through a consumption survey or wealth index.

Indicator	What it measures	Population	Data collection	Data analysis	Notes
Wealth indices / poverty levels	Wealth / socioeconomic status, a proxy for income.	Household	Various methodologies exist (see notes), all of which are based on a household survey		The DHS contains a wealth index. Poverty rates are usually monitored by Governments. A gender-sensitive indicator guide is available online.[53]
Sales of agricultural products	Value of incremental sales (collected at farm-level) attributed to project implementation.	Household	Household survey and/or Enterprise records		USAID uses the indicator "value of incremental sales (collected at farm-level), attributed to Feed the Future implementation" found in FTF, 2016.[31]
Income or consumption		Household	Household survey and/or Enterprise records. A detailed household consumption survey would typically not be undertaken by a single project, but is rather part of Household Consumption and Expenditure Surveys administered periodically in most countries (including Living Standards and Measurement Studies (LSMS), Household Budget Surveys (HBS), etc.).		There are various ways this indicator could be defined.* The majority of investment projects in agriculture, rural development and value chain expect to increase incomes and aim to demonstrate these at design stage by undertaking an Economic and Financial Analysis (EFA) of the project based on crop budgets, farm models and enterprise models. The EFA guidelines under development by IFAD (with contribution by the FAO Investment Center) can be used as a reference when finalized. The first volume (basic concepts and rationale) is already available online.[54] In 2016, two more volumes should be published, the last one comprising a series of case studies, including one on nutrition-sensitive agriculture investment. Such projected increased incomes should then be monitored during project implementation.

Indicator	What it measures	Population	Data collection	Data analysis	Notes
Household asset index	The sets of key assets within the household	Household	Household asset lists can be gathered as part of a household survey.	The set of key assets can change from one rural context to another; the final composition of the asset list should reflect distinct consumer preferences. Once the list is compiled, a monetary unit values is attributed to each of the assets, then the index is calculated as the total value of all assets owned by the household.	The assumption underlying this indicator is that households with a greater investment in key consumer durables are more economically secure, i.e. they have access to more income. An asset index is part of a review published by the Livelihood Monitoring Unit (LMU) Rural Livelihoods Program CARE Bangladesh, *Measuring Livelihood Impacts: A Review of Livelihoods Indicators.*[55]

* Examples could be: "increased farm and off-farm incomes (including from micro-enterprises promoted by then project) as a result of a project" or "prevalence of households raised above the poverty line as a result of the project."

Table 7.6 Women's empowerment

- **When to use**: some assessment of gendered impact on income and time should be undertaken to ensure no harm and equity.
- Given that validated indicators are unavailable, assessment may be qualitative.
- Women's empowerment has several aspects: income, time/labour, assets, knowledge, decision-making, etc. Each may be affected differently or more/less strongly by the intervention. It is important to measure the aspects most likely to be affected by the intervention.

Indicator	What it measures	Population	Data collection	Data analysis	Notes
Women's Empowerment in Agriculture Index (WEAI)	A composite measurement tool that indicates women's control over critical parts of their lives in the household, community and economy	Women	Household survey	It measures five domains (i) decisions about agricultural production; (ii) access to and decision-making power over productive resources; (iii) control over use of income; (iv) leadership in the community; and (v) time use. It also measures women's empowerment relative to men within their households. Some components of the index may be more likely to change as a result of intervention than others. The components of the index can be presented separately, in terms of proportion of women not empowered in each domain. A comparison to men's scores shows gender gaps in empowerment.	The WEAI can identify women who are disempowered and understand how to increase autonomy. **METHODOLOGY** - IFPRI Women's Empowerment in Agriculture Index, 2012;[56] - *Instructional Guide on the Women's Empowerment in Agriculture Index*, Alkire *et al.*, 2013.[57]

Indicator	What it measures	Population	Data collection	Data analysis	Notes
Women's control of income	The extent to which women control decisions around how income is used. Methodology is not standardized	Women	Household survey; women should be the respondents	Collection of data on decision-making concerning the use of proceeds from farm plots (it can be modified to include proceeds from other income generating activities).	**METHODOLOGY** Agri-gender statistics toolkit: income and expenditure questionnaire.[58]
Women's time use and labour	Percentage of time spent daily in household on paid and nonpaid activities, disaggregated. Division of labour and responsibility within the household		Detailed methodology from national Time Use Surveys, or a simplified methodology using either a time diary or a 24-hour recall. This information may also be obtained through qualitative methods such as focus group discussions	Time use data are analysed by categories of time use (e.g. agricultural work, leisure, child care, etc.)	May be useful to ensure project is not creating unwanted time burdens for women. In a 24-hour recall method (used in the WEAI), respondents do not keep their own time diary, but are rather asked how they spent their time the previous day. The basic objective of a time diary method is to enable respondents to report all activities undertaken over a prescribed period of time and the beginning and ending time for each activity. There are two basic types of diaries: the full time diary and the simplified time diary. Direct observation method: the time use of the respondent is observed and recorded by the survey enumerator. Observation can be carried out on a continuous basis or on a random spot basis. **METHODOLOGIES** WEAI time module available from IFPRI *Women's Empowerment in Agriculture Index*, 2012.[56] Agri-gender statistics toolkit: labour and time use questionnaire.[59] Description of data collection methods available here: *Guide to producing statistics on time use: measuring paid and unpaid work*. UN Economic and Social Affairs, 2005.[60] Some further information is available online (United Nations Gender statistics).[61]

Indicator	What it measures	Population	Data collection	Data analysis	Notes
Asset ownership by gender	It measures access to productive resources such as (i) land and water; (ii) farm inputs; (iii) farm implements, assets and technologies; (iv) credit; (v) extension services and training programmes	Women	Household survey; women should be the respondents		**METHODOLOGIES** Examples of already formulated questionnaires and questions are available in the FAO Agri-gender statistics toolkit: questionnaire on access to productive resources.[62]
Qualitative process to understand equity, time use and income control	Women's empowerment	Women	Focus groups, interviews, observation		**METHODOLOGIES** Qualitative inquiry can take many forms, but two guides have recently been developed to understand gender equity qualitatively: - CARE *Gender Toolkit*[6;3] - Land O'Lakes *Integrating Gender throughout a Project's Life Cycle 2.0*, 2015.[64]

Additional resources have been published by Alkire *et al.*, 2013[65], Malapit *et al.*, 2014[66], the World Bank and FAO, 2009.[67]

Table 7.7 Nutrition and food safety knowledge and norms

- **When to use**: when intervention is promoting certain nutrition behaviours or messages; or, to understand likelihood of consumption of specific foods or overall dietary patterns for various population sub-groups.

Indicator	What it measures	Population	Data collection	Data analysis	Notes
Indicator of nutrition and food safety-related knowledge – to be specified according to intervention	Nutrition and food safety-related knowledge and attitudes (KAP) at the community level	Usually women	Household survey and/or qualitative process		These indicators will be project-specific, depending on what sort of knowledge or behaviour is promoted. **VALIDITY** Knowledge and attitudes do not refer to physical objects but to psychosocial and subjective concepts. It is therefore not possible to validate the results concerning knowledge and attitudes in KAP surveys because no objective benchmark or reference exists. **METHODOLOGY** (standardized) FAO *Guidelines for assessing nutrition-related Knowledge, Attitudes and Practices* (2014)[17] comprise predefined questionnaires that capture information on critical knowledge, attitudes and practices related to most common nutrition topics. Note: if agricultural knowledge (e.g. knowledge of improved practices) is sometimes assessed in projects, relevant nutritional knowledge could be added.
Changes in specific behaviours promoted with regard to food safety	Awareness about safety at household (consumers') level	Households or community	Household survey and/or qualitative process		Indicators would be intervention-specific. They could also be built around the concept of the WHO 5 keys for safer foods.[68]

Table 7.8 Care practices

- **When to use**: when intervention is promoting certain nutrition behaviours or messages; or, to understand likelihood of consumption of specific foods or overall dietary patterns for various population sub-groups.

Indicator	What it measures	Population	Data collection	Data analysis	Notes
Breastfeeding indicators	Frequency, duration, or completeness of breastfeeding	Children under 2 years (mainly)	Recall of the previous day, administered through a household survey		There are several indicators of breastfeeding defined in the WHO indicator guide referenced. **VALIDITY** These indicators are very useful for capturing feeding practices below 2 years of age, but are not validated against anything but themselves. **METHODOLOGY** Indicators for assessing infant and young child feeding practices (WHO, 2008).[11,12] These may be useful if the project includes a nutrition education component focused on infant feeding, or to ensure that no harm is being done to women's time/ ability to breastfeed.
Minimum Acceptable Diet (MAD)	This indicator combines standards of (i) dietary diversity (a proxy for nutrient density); and (ii) feeding frequency (a proxy for energy density) by breastfeeding status; and thus provides a useful way to track progress at simultaneously improving the key quality and quantity dimensions of children's diets	Children under 2 years	Recall of the previous day, administered through a household survey	This is a composite indicator: while it is an indicator of diet quality for young children, it is primarily an indicator of care practices, since those determine young child diet quality to such a large extent. Can be used to calculate the proportion of children 6-23 months of age who receive a MAD	**VALIDITY** Validation studies have been done on the minimum dietary diversity component (see diet quality section), but not on the composite indicator. **METHODOLOGY** Indicators for assessing infant and young child feeding practices - Minimum Acceptable Diet, published by WHO, 2008.[11,12]

Indicator	What it measures	Population	Data collection	Data analysis	Notes
Minimum Meal Frequency	Proxy for energy intake from non-breastmilk foods among young children	Children under 2 years	Recall of the previous day, administered through a household survey	Proportion of breastfed and non-breastfed children 6–23 months of age who receive solid, semi-solid, or soft foods (but also including milk feeds for non-breastfed children) a minimum number of times or more	**CUTOFF** Minimum is defined as: – 2 times for breastfed infants 6–8 months; – 3 times for breastfed children 9–23 months; – 4 times for non-breastfed children 6–23 months; – "Meals" include both meals and snacks (other than trivial amounts) and frequency is based on caregiver report. **METHODOLOGY** Indicators for assessing infant and young child feeding practices - Minimum Meal Frequency, published by WHO, 2008.[11,12]

Minimum Dietary Diversity (children age 6-23 months) (full description available in Table 7.1).[11,12]

Additional resources:

• If early child development is an important aim or results from collaboration with other sectors or projects, proxy indicators for early childhood care and education may be of interest. Two types of indicators include:

  – Home Observations for Measurement of the Environment (HOME): combination of interview and direct observation. Interviewers provide specific time limits as a framework for the conversation by asking the caregiver to focus on the facts of a very specific day of the week. The HOME takes 45–60 minutes to administer and requires skilled, well-trained interviewers and considerable adaptation when used in developing countries. Moreover, the HOME involves observations, which are more difficult to standardize. There is no standardized procedure for administration; information is obtained by only one informant each time on only one occasion, which might be unrepresentative of a child's full life conditions (Totsika and Sylva, 2004[69] and Iltus, 2006[70]).

  – Family Care Indicators (FCI): developed to measure home stimulation in large populations and were derived from the HOME indicators. The FCI questionnaire was developed by groups of experts organized by UNICEF with preliminary piloting for comprehension in several countries (Hamadani et al., 2010[71]).

Table 7.9 Natural resource management practices, health and sanitation environment*

- **When to use**: when intervention affects soil or water management, or livestock-human interactions.
- These indicators will be project-specific, depending on what area of natural resources or health environment that the agricultural activities may affect.
- The dimensions of the health and sanitation environment most relevant to agriculture interventions could include water quantity and quality, environmental contamination having an impact on food safety, agrochemical exposure, risk of zoonotic or water vector-borne disease and cleanliness of children's play areas (presence of animals in or near the home).

Indicator	What it measures	Population	Data collection	Data analysis	Notes
Access to improved drinking water source	See indicator definitions	Household	Household survey		**DEFINITION** The following specific indicators have been defined: (1) percentage of population using an improved drinking water source on premises with discontinuity less than 2 days in the last 2 weeks, with less than 10 cfu E.coli / 100ml year round at source, accessible to all members of the household at the times they need it; (2) percentage of population using an improved water source with a total collection time of 30 minutes or less for a roundtrip including queuing. The WHO/UNICEF Joint Monitoring Programme has established a standard set of drinking-water and sanitation categories that are used for monitoring purposes.[72,73]
Presence of animals in/ near household	Indicates risk of environmental enteropathy	Household	Household survey		A specific indicator and methodology is not defined.
Sustainability of water availability and water use efficiency measures	See notes	Children under 5	Household survey		Possible indicators would depend on project context and interventions. They could include: -percentage of delivered vs. required water; -number of farmers with secure access to water. These are sample indicators from IFAD's *Results and Impact Management System (RIMS) handbook* (2014).[7]

Indicator	What it measures	Population	Data collection	Data analysis	Notes
Nutrition indicators for biodiversity	Indicates sub-species/ varietal diversity of foods consumed	Household or individual	Household survey		**DEFINITION** The indicator is a count of the number of foods consumed with a sufficiently detailed description to identify genus, species, subspecies and variety / cultivar / breed and with at least one value for a nutrient or other bioactive component. FAO Nutrition Indicators for Biodiversity are available online, 2008.[74,75]
Contamination from water or environment in food supply		Household or community			Possible indicators would depend on project context and interventions. They could be related to: - water quality to be used in food production (from primary production to consumers); - contamination of soils (natural, industrial); - adoption of mitigating practices by farmers/producers (modification in agricultural practices; change of use of soils); - percentage of wastewater being treated/produced (this is an indicator used as part of SDG 6.3 measurement).

*Project managers may wish to consult standard sanitation indicators to understand the health environment, even though the agriculture/food investment may not affect these indicators. The WHO/UNICEF Joint Monitoring Programme has established a standard set of drinking-water and sanitation categories that are used for monitoring purposes.[72,73] These include:

- Access to basic hand-washing facilities in the home:
  - percentage of households with soap and water at a hand washing facility commonly used by family members;
  - percentage of households with soap and water at a hand-washing facility within or immediately near sanitation facilities;
  - percentage of households with soap and water at a hand-washing facility within or immediately near the food preparation area.
- Access to adequate sanitation facilities:
  - percentage of population using an adequate sanitation facility.

Table 7.10 Nutritional status: anthropometric indicators

- Note: as detailed above, these indicators are often insensitive to short-term change in agriculture projects.
- Further info on child growth indicators and their interpretation available on WHO website.[76]
- Reference population of the corresponding age are established by the WHO's child growth standards available on WHO website.[77]

Indicator	What it measures	Population	Data collection	Data analysis	Notes
Stunting	Height for age	Children under 5	Household survey	<-2 Z scores is the cutoff for moderate level; <-3 Z scores is the cutoff for severe level	Requires carrying height boards to measure heights of children and specific training for accurate measurement. Requires determining child's age in months accurately. Would usually not allow to show observable changes in many small-scale interventions and over short periods of time.
Wasting	Weight for height	Children under 5	Household survey	<-2 Z scores is the cutoff for moderate level; <-3 Z scores is the cutoff for severe level	Requires carrying height boards and weighing scales to measure heights and weights.
Underweight	Weight for age	Children under 5	Household survey	<-2 Z scores is the cutoff for moderate level; <-3 Z scores is the cutoff for severe level	Requires carrying scales to measure weights of children; requires determining child's age in months accurately.
Maternal weight/BMI	Weight in kg/height in m2	Usually adult women	Household survey	<18.5 is the cut-off for underweight; >25 is the cut-off for overweight for many countries; >30 is the cut-off for obesity	Requires carrying scales to measure weights of women.

Additional resources have been published by Cogill, 2003[78], United Nations, 1986[79], UNICEF *Harmonized training package for nutrition. Measuring undernutrition in individuals*.[80]

Useful considerations and techniques for anthropometric measurement can be found in IFAD's *Practical guidance for baseline, mid-term and impact surveys*.[81,82]

Table 7.11 Nutritional status: biochemical indicators

- Note: as detailed above, these indicators are not appropriate for all projects, as they are more expensive and invasive than other outcomes to measure and many projects are not designed to affect them in the short term.
- This table includes indicators of nutrients for which many populations are deficient and that meet *both* of two criteria: (i) they may be affected by the foods made available through agricultural activities; and (ii) they can be measured with reasonable precision and cost at an individual level. Other important micronutrients (iodine, zinc, vitamin B12) do not fit those criteria.

Indicator	What it measures	Population	Data collection	Data analysis	Notes
Iron status	Whether an individual's body is deficient or replete in iron	Usually women or children under 5	Requires collecting blood for 3-4 different tests of iron biomarkers and usually also requires tests for inflammation.		Assessing the iron status of populations: report of a joint WHO/Centers for Disease Control and Prevention technical consultation (WHO and CDC, 2007).[83]
Anaemia	Haemoglobin level	Usually women or children under 5	Blood samples	Compare data to WHO universal thresholds that define levels of public health importance.	Document to assess haemoglobin concentrations for the diagnosis of anaemia and assessment of severity available on WHO website (WHO, 2011).[84]
Vitamin A status	Whether an individual's body is deficient or replete in vitamin A	Usually women or children under 5	Clinical signs (Bitot's spots, xerophthalmia); blood collection; breastmilk collection. Usually also requires tests for inflammation.	Serum retinol or breastmilk retinol.	Reference document for assessing vitamin A deficiency in monitoring and evaluating interventions, available on WHO website (WHO, 1996).[85]

Additional detailed information available in: Gibson, 2005.[22]

Other important micronutrients include iodine, zinc and vitamin B12. While agriculture projects may influence consumption of zinc-rich and B12-rich foods, which may translate into improved nutrient status, these nutrients are quite difficult and costly to measure on an individual basis. Iodine is not usually influenced by agriculture (except in the case of iodine fertilization of soils). More information is available here:

• Zinc: plasma or serum zinc levels are the most commonly used indices for evaluating zinc deficiency, but these levels do not necessarily reflect cellular zinc status due to tight homeostatic control mechanisms (Prasad, 1985);[86]

• Vitamin B12 is assessed by blood samples. More information can be found on FAO website;[87]

• Iodine status is usually tested through urine samples (clinical signs: goitre and impaired mental functions). Even if populations may have attained iodine sufficiency by median urinary iodine concentration, goitre may persist, even in children. Reference documents:

  – *Assessment of Iodine Deficiency Disorders and Monitoring their Elimination* (third edition), WHO, 2007;[88]

  – *Food and Nutrition in Numbers 2014*. FAO, 2014.[89]

References

1 FAO/WHO. 2014a. *Rome declaration on Nutrition*. Outcome of the Second International Conference on Nutrition Rome, 19-21 November 2014, Rome, FAO. Available at: www.fao.org/3/a-ml542e.pdf

2 FAO/WHO. 2014b. *Framework for Action*. Outcome of the Second International Conference on Nutrition Rome, 19-21 November 2014, Rome, FAO. Available at: www.fao.org/3/a-mm215e.pdf

3 United Nations. 2015. *Millennium Development Goals Report, 2015*. New York, USA, United Nations. Available at: www.un.org/millenniumgoals/2015_MDG_Report/pdf/MDG%202015%20rev%20(July%201).pdf

4 FAO. 2015. *Key recommendations for improving nutrition through agriculture and food systems*. Rome, FAO. Available at: www.fao.org/3/a-i4922e.pdf

5 Herforth, A., Dufour, C. & Noack, A.L. 2015. *Designing nutrition-sensitive agriculture investments: checklist and guidance for programme formulation*. Rome, FAO. Available at www.fao.org/3/a-i5107e.pdf

6 Herforth, A. & Ballard, T. 2016. Nutrition indicators in agriculture projects: current measurements, priorities and gaps. *Global Food Security*. Available at: www.sciencedirect.com/science/article/pii/S22119123415300109 (accessed 25.08.2016)

7 IFAD. 2014. *Results and Impact Management System (RIMS) handbook*. Rome, IFAD. Available at: www.ifad.org/documents/10180/9c36cfc5-28d3-401e-b30c-acec8d6acd00 (accessed 29.08.2016)

8 International Household Survey Network. [Website] Available at: www.ihsn.org/home/ (accessed 30.08.2016)

9 Herforth, A. & Ahmed, S. 2015. The food environment, its effects on dietary consumption and potential for measurement within agriculture-nutrition interventions. *Food Security* 7(3): 505-520. Available at: http://link.springer.com/article/10.1007/s12571-015-0455-8 (accessed 25.08.2016)

10 FAO/FHI 360. 2016. *Minimum Dietary Diversity for Women: a Guide for Measurement*. Rome, FAO. Available at: www.fao.org/3/a-i5486e.pdf

11 WHO. 2008. *Indicators for assessing infant and young child feeding practices: conclusions of a consensus meeting held 6–8 November 2007 in Washington D.C.* Geneva, WHO. Available at: www.who.int/maternal_child_adolescent/documents/9789241596664 (accessed 25.08.2016)

12 WHO. 2010. *Indicators for assessing infant and young child feeding practices. Part 2-Measurement*. Geneva, WHO. Available at: www.who.int/nutrition/publications/infantfeeding/9789241599290 (accessed 09.09.2016)

13 FAO Voices of the Hungry project. Available at: www.fao.org/in-action/voices-of-the-hungry/en/#.V8WB6Ft96Uk (accessed 30.08.2016)

14 Shannon index- Biodiversity calculator. Available at: www.alyoung.com/labs/biodiversity_calculator.html (accessed 25.08.2016)

15 Simpson index- Biodiversity calculator. Available at: www.alyoung.com/labs/biodiversity_calculator.html (accessed 25.08.2016)

16 Bilinsky, P. & Swindale, A. 2010. *Months of Adequate Household Food Provisioning (MAHFP) for Measurement of Household Food Access: Indicator Guide* (v.4). Washington, D.C., FHI 360/FANTA. Available at: www.fantaproject.org/sites/default/files/resources/MAHFP_June_2010_ENGLISH_v4.pdf

17 Fautsch Macías, Y. & Glasauer, P. 2014. *Guidelines for assessing nutrition-related Knowledge, Attitudes and Practices*. Rome, FAO. Available at: www.fao.org/3/a-i3545e.pdf

18 vWHO/UNICEF. Joint Monitoring Programme standard set of drinking-water and sanitation categories. Available at: www.wssinfo.org/definitions-methods/watsan-categories (accessed 25.08.2016)

19 Lele, U., Masters, W. A., Kinabo, J. Meenakshi J.V., Ramaswami, B., Tagwireyi, J., Bell W.F.L. & Goswami, S. 2016. *Food Security and Nutrition: An Independent Technical Assessment and User's Guide for Existing Indicators*. Rome, FAO, Washington DC, IFPRI and Rome, WFP. Available at: www.fao.org/fileadmin/user_upload/fsin/docs/1_FSIN-TWG_UsersGuide_12June2016.compressed.pdf

20 Swindale, A. & Bilinsky, P. 2006. *Household Dietary Diversity Score (HDDS) for Measurement of Household Food Access: Indicator Guide* (v.2). Washington, D.C., FHI 360/FANTA. Available at: www.fantaproject.org/sites/default/files/resources/HDDS_v2_Sep06_0.pdf

21 Torheim, L.E., Barikmo, I, Parr, C.L., Hatloy, A., Ouattara, F. & Oshaug, A. 2003. Validation of food variety as an indicator of diet quality assessed with a food frequency questionnaire for Western Mali. *Eur J Clin Nutr* 57: 1283–1291. Available at: www.nature.com/ejcn/journal/v57/n10/pdf/1601686a.pdf

22 Gibson, R. S. (2005). *Principles of Nutritional Assessment* (second edition). Oxford, Oxford University Press

23 Gibson, R. S. & Ferguson, E. L. 2008. *An interactive 24-hour recall for assessing the adequacy of iron and zinc intakes in developing countries*, HarvestPlus Technical Monograph 8. Washington, DC, IFPRI and Cali, International Center for Tropical Agriculture (CIAT). Available at: www.ifpri.org/sites/default/files/publications/tech08.pdf

24 CGIAR. 2014. CGIAR Research programme on Agriculture for Nutrition and Health. Extension proposal 2015-2016 submitted to the CGIAR consortium board, April 2014. Available at: http://a4nh.cgiar.org/files/2014/03/A4NH-Extension-Proposal-2015-2016FINAL.pdf

25 FAO. 2015. *Guidelines on the collection of information on food processing through food consumption surveys*. Rome, FAO. Available at: www.fao.org/3/a-i4690e.pdf

26 Monteiro, C. A., Moubarac, J.C., Cannon, G., Popkin, S. W. Ng, B. 2013. Ultra-processed products are becoming dominant in the global food system. *Obesity reviews* 14 (2): 21-28. Available at: http://onlinelibrary.wiley.com/doi/10.1111/obr.12107/full (accessed 30.08.2016)

27 Mubarac, J.C., Batal, M., Martins, A. P., Claro, R. 2014. Processed and ultra-processed food products: consumption trends in Canada from 1938 to 2011. *Can J Diet Pract Res. 2014 Spring*; 75(1):15-21. Abstract available at: www.ncbi.nlm.nih.gov/pubmed/24606955 (accessed 25.08.2016)

28 Persson, V., Greiner, T., Bhagwat, I.P. & Gebre-Medhin, M. 1998. The Helen Keller international food frequency method may underestimate vitamin A intake where milk is a normal part of the young child diet. *Ecology Of Food And Nutrition* 8: 67-69. Available at: www.researchgate.net/publication/254230027_The_Helen_Keller_International_Food_Frequency_Method_may_underestimate_vitamin_A_intake_where_milk_is_a_normal_part_of_the_young_child_diet (accessed 30.08.2016)

29 Codex Alimentarius. *Guidelines on nutrition labelling* (Rev 2013 and 2015). Rome, FAO, Geneva, WHO. Available at: www.fao.org/input/download/standards/34/CXG_002e_2015.pdf

30 Codex Alimentarius. *Guidelines for use of nutrition and health claims*. 1997. Rome, FAO, Geneva, WHO. Available at: www.fao.org/input/download/standards/351/CXG_023e.pdf

31 Feed the Future. 2016. *Indicator Handbook: Definition Sheets*. Updated 2016. Working document describing the indicators selected for monitoring and evaluation of the U.S. Government's global hunger and food security initiative, Feed the Future. Washington, DC. Feed the Future. Available at: https://feedthefuture.gov/sites/default/files/resource/files/Feed_the_Future_Indicator_Handbook_25_July_2016.pdf

32 FAO. 2012a. *Guidelines for measuring household and individual dietary diversity*. Rome, FAO. Available at: www.fao.org/docrep/014/i1983e/i1983e00.htm (accessed 26.08.2016)

33 WFP-VAM. 2008. *Technical Guidance Sheet - Food Consumption Analysis: Calculation and Use of the Food Consumption Score in Food Security Analysis*. Available at: www.wfp.org/content/technical-guidance-sheet-food-consumption-analysis-calculation-and-use-food-consumption-score-food-s (accessed 26.08.2016)

34 Coates, J., Swindale A. & Bilinsky, P. 2007. *Household Food Insecurity Access Scale (HFIAS) for Measurement of Household Food Access: Indicator Guide* (v. 3). Washington, D.C., Food and Nutrition Technical Assistance Project (FANTA), Academy for Educational Development. Available at: www.fao.org/fileadmin/user_upload/eufao-fsi4dm/doc-training/hfias.pdf

35 FAO. 2012b. *Escala Latinoamericana y Caribeña de Seguridad Alimentaria (ELCSA): Manual de uso y aplicación*. Rome, FAO. Available at: www.fao.org/docrep/019/i3065s/i3065s.pdf

36 Ballard, T., Coates, J., Swindale, A. & Deitchler, M. 2011. *Household Hunger Scale: Indicator Definition and Measurement Guide*. Washington, DC, Food and Nutrition Technical Assistance II (FANTA II) Project, FHI 360. Available at: www.fantaproject.org/monitoring-and-evaluation/household-hunger-scale-hhs (accessed 26.08.2016)

37 WFP-VAM. Coping Strategies Index (CSI). Available at: http://resources.vam.wfp.org/node/6 (accessed 26.08.2016)

38 Jones, A.D., Ngure, F. M., Pelto, G. & Young, S. L. 2013. What are we assessing when we measure food security? A compendium and review of current metrics. *Adv Nutr.* 4: 481-505. Available at: www.fao.org/fileadmin/templates/ess/documents/meetings_and_workshops/cfs40/001_What_Are_We_Assessing_When_We_Measure_Food_Secuirty.pdf

39 Coates, J. 2013. Build it back better: Deconstructing food security for improved measurement and action. *Global Food Security* 2(3):188–194

40 FAO/WFP. 2012. *Household Dietary Diversity Score and Food Consumption Score: A joint statement*. Available at http://documents.wfp.org/stellent/groups/public/documents/ena/wfp269531.pdf

41 Remans, R., Flynn, D.F.B., DeClerck, F., Diru, W., Fanzo, J., Gaynor, K., Lambrecht, I., Mudiope, J., Mutuo, P.K., Nkhoma, P., Siriri, D., Sullivan, C. & Palm, C.A. 2011. Assessing Nutritional Diversity of Cropping Systems in African Villages. *PLoS ONE* 6(6). Available at: http://journals.plos.org/plosone/article?id=10.1371%2Fjournal.pone.0021235 (accessed 26.08.2016)

42 Global Strategy to improve Agricultural and Rural Statistics (GSARS). [Website] Available at: http://gsars.org/en/ (accessed 26.08.2016)

43 FAO Food Price Monitoring and Analysis. [Website] Available at: www.fao.org/giews/food-prices/home (accessed 29.08.2016)

44 WFP- VAM Food and Commodity Prices Data Store. [Website] Available at: http://foodprices.vam.wfp.org/ (accessed 29.08.2016)

45 Save the Children The Minimum Cost of A Healthy Diet. [Website] Available at: www.savethechildren.org.uk/resources/online-library/the-minimum-cost-of-a-healthy-diet (accessed 29.08.2016)

46 USAID Cost of the Diet (CoD). [Website] Available at: www.spring-nutrition.org/publications/tool-summaries/cost-diet (accessed 29.08.2016)

47 FAO/WHO. 2013. *Histamine Sampling Tool User Guide* (Version 1.0). Rome, FAO. Available at: www.fstools.org/histamine (accessed 29.08.2016)

48 FAO. 2014. *Mycotoxin sampling Tool User Guide* (Version 1.0 and 1.1). Rome, FAO. Available at: www.fstools.org/mycotoxins/Documents/UserGuide.pdf

49 FAO Save Food: Global Initiative on Food Loss and Waste Reduction- Field Case Studies. [Website] Available at: www.fao.org/save-food/resources/casestudies/en/ (accessed 08.09.2016)

50 FAO Food Safety and Quality. [Website] Available at: www.fao.org/food/food-safety-quality/home-page/en/(accessed 29.08.2016)

51 FAO. 1998. *Food Quality and Safety Systems - A Training Manual on Food Hygiene and the Hazard Analysis and Critical Control Point (HACCP) System.* Rome, FAO. Available at: www.fao.org/docrep/w8088e/w8088e04.htm (accessed 29.08.2016)

52 Codex Alimentarius. *Standards, guidelines and advisory texts.* Available at: www.fao.org/fao-who-codexalimentarius/standards/en/ (accessed 29.08.2016)

53 Njuki, J., Poole, J., Johnson, N., Baltenweck,I., Pali, P., Lokman, Z., & and Mburu, S. 2011. *Gender, Livestock and Livelihood Indicators.* Nairobi, ILRI. Available at: https://cgspace.cgiar.org/bitstream/handle/10568/3036/Gender%20Livestock%20and%20Livelihood%20Indicators.pdf

54 IFAD. 2015. *Economic and Financial Analysis of rural investment projects-Basic Concepts and rationale.* Rome, IFAD. Available at: www.ifad.org/documents/10180/a53a6800-7fab-4661-ac78-faefcb7f00f8 (accessed 29.08.2016)

55 CARE Bangladesh Livelihood Monitoring Unit (LMU). 2004. *Measuring Livelihood Impacts: A Review of Livelihoods Indicators.* Rural Livelihoods Program. Available at: http://portals.wi.wur.nl/files/docs/ppme/LMP_Indicators.pdf

56 IFPRI. 2012. *Women's Empowerment in Agriculture Index.* Washington, D.C., IFPRI. Available at: www.ifpri.org/publication/womens-empowerment-agriculture-index (accessed 29.08.2016)

57 Alkire, S., Malapit, H., Meinzen-Dick, R., Peterman, A., Quisumbing, A. R., Seymour, G. & Vaz, A. 2013. *Instructional Guide on the Women's Empowerment in Agriculture Index.* Washington, DC., IFPRI. Available at: www.ifpri.org/sites/default/files/Basic%20Page/weai_instructionalguide_1.pdf

58 FAO-Agri-gender statistics toolkit: income and expenditures questionnaire. Available at: www.fao.org/fileadmin/templates/gender/agrigender_docs/q6.pdf

59 FAO-Agri-gender statistics toolkit: labour and time use questionnaire. Available at: www.fao.org/fileadmin/templates/gender/agrigender_docs/q5.pdf

60 United Nations Economic and Social Affairs (UNDESA). 2005. *Guide to producing statistics on time use: measuring paid and unpaid work.* New York, USA, United Nations. Available at: http://unstats.un.org/unsd/publication/SeriesF/SeriesF_93e.pdf

61 United Nations Gender Statistics. *Allocation of time and time-use.* Available at: http://unstats.un.org/unsd/gender/timeuse/ (accessed 29.08.2016)

62 FAO-Agri-gender statistics toolkit: access to productive resources questionnaire. Available at: www.fao.org/fileadmin/templates/gender/agrigender_docs/q2.pdf

63 CARE Gender Toolkit. Available at: http://gendertoolkit.care.org/default.aspx (accessed 29.08.2016)

64 Land O'Lakes. 2015. *Integrating Gender throughout a Project's Life Cycle 2.0. A Guidance Document for International Development Organizations and Practitioners.* Washington, D.C., Land O'Lakes. Available at: www.landolakes.org/resources/tools/Integrating-Gender-into-Land-O-Lakes-Technical-App (accessed 29.08.2016)

65 Alkire, S., Meinzen-Dick, R., Peterman, A., Quisumbing, A., Seymour, G. & Vaz, A. 2013. *The Women's Empowerment in Agriculture Index.* OPHI Working Paper No. 58

66 Malapit, H.J., Sproule, K., Kovarik, C., Meinzen-Dick, R., Quisumbing, A., Ramzan, F., Hogue, E. & Alkire, S. 2014. *Women's Empowerment in Agriculture Index: Baseline Report.* Washington, D.C., IFPRI. Available at: https://feedthefuture.gov/sites/default/files/resource/files/ftf_progress_weai_baselinereport_may2014.pdf

67 World Bank/FAO. 2009. *Gender in Agriculture Sourcebook.* Washington, DC., World Bank. Available at: http://siteresources.worldbank.org/INTGENAGRLIVSOUBOOK/Resources/CompleteBook.pdf

68 WHO. 2006. *Five keys to safer food manual.* Geneva, WHO. Available at: www.who.int/foodsafety/publications/5keysmanual (accessed 29.08.2016)

69 Totsika, V. & Sylva, K. 2004. The home observation for measurement of the environment revisited. *Child and Adolescent Mental Health,* 9 (1): 25–35

70 Iltus, S. 2006. Significance of home environments as proxy indicators for early childhood care and education. Background paper prepared for the Education for *All Global Monitoring Report 2007 Strong foundations: early childhood care and education.* New York, USA, UNESCO. Available at: http://unesdoc.unesco.org/images/0014/001474/147465e.pdf

71 Hamadani, J.D., Tofail, F., Hilaly, A., Huda, S.N., Engle, P. & Grantham-McGregor, S.M. 2010. Use of Family Care Indicators and Their Relationship with Child Development in Bangladesh. *J Health Popul Nutr* 28 (1): 23–33. Available at: www.ncbi.nlm.nih.gov/pmc/articles/PMC2975843/ (accessed 29.08.2016)

72 WHO/UNICEF Joint Monitoring Programme (JMP) for water supply and sanitation. [Website] Available at: www.wssinfo.org (accessed 29.08.2016)

73 WHO/UNICEF Joint Monitoring Programme (JMP) for water supply and sanitation- standard set of drinking-water and sanitation categories for monitoring purposes. [Website] Available at: www.wssinfo.org/definitions-methods/watsan-categories (accessed 29.08.2016)

74 FAO. 2008a. *Expert Consultation on Nutrition Indicators for Biodiversity-food composition.* Rome, FAO. Available at: www.fao.org/3/a-a1582e.pdf

75 FAO. 2008b. *Expert Consultation on Nutrition Indicators for Biodiversity-food consumption.* Rome, FAO. Available at: www.fao.org/docrep/014/i1951e/i1951e00.htm (accessed 29.08.2016)

76 WHO Global Database on Child Growth and Malnutrition- Child growth indicators and their interpretation. [Website] Available at: www.who.int/nutgrowthdb/about/introduction/en/index2.html (accessed 29.08.2016)

77 WHO. 2006. *WHO child growth standards: length/height-for-age, weight-for-age, weight-for-length, weight-for height and body mass index-for-age: methods and development.* Geneva, WHO. Available at: www.who.int/childgrowth/standards/Technical_report.pdf

78 Cogill, B. 2003. *Anthropometric Indicators Measurement Guide.* Washington, DC, Food and Nutrition Technical Assistance (FANTA) Project, FHI 360. Available at: www.fantaproject.org/sites/default/files/resources/anthropometry-2003-ENG.pdf

79 United Nations. 1986. *How to Weigh and Measure Children: Assessing the Nutritional Status of Young Children in Household Surveys*. New York, USA, United Nations. Available at: http://unstats.un.org/unsd/publication/unint/dp_un_int_81_041_6E.pdf

80 UNICEF. *Harmonized training package for nutrition. Measuring undernutrition in individuals*. [Online training] Available at: www.unicef.org/nutrition/training/3.1/1.html (accessed 29.08.2016)

81 IFAD. 2005a. *Practical guidance for impact surveys*. Rome, IFAD. Available at: www.ifad.org/documents/10180/78da2b7e-9b3a-4f98-b514-2783a85234a2 (accessed 29.08.2016)

82 IFAD. 2005b. *Practical guidance for impact surveys- tools for conducting an impact survey*. Rome, IFAD. Available at: www.ifad.org/operations/rims/guide/e/part2_e.pdf

83 WHO/CDC, 2007. *Assessing the iron status of populations: report of a joint WHO/ Centers for Disease Control and Prevention technical consultation on the assessment of iron status at the population level*, 2nd ed. Geneva, WHO. Available at: www.who.int/nutrition/publications/micronutrients/anaemia_iron_deficiency/9789241596107/en/ (accessed 29.08.2016)

84 WHO. 2011. *Haemoglobin concentrations for the diagnosis of anaemia and assessment of severity*. Geneva, WHO. Available at: www.who.int/vmnis/indicators/haemoglobin/en (accessed 29.08.2016)

85 WHO. 1996. *Indicators for assessing vitamin A deficiency and their application in monitoring and evaluating intervention programmes*. Geneva, WHO. Available at: www.who.int/nutrition/publications/micronutrients/vitamin_a_deficiency/WHO_NUT_96.10/en/ (accessed 29.08.2016)

86 Prasad, A. S. 1985. Laboratory diagnosis of zinc deficiency. *J Am Coll Nutr.* 4 (6):591-8. Abstract available at: www.ncbi.nlm.nih.gov/pubmed/4078198 (accessed 29.08.2016)

87 WHO/FAO. 2004. *Vitamin and mineral requirements in human nutrition* (second edition). Geneva, WHO. Available at: http://apps.who.int/iris/bitstream/10665/42716/1/9241546123.pdf?ua=1 (accessed 09.09.2016)

88 WHO. 2007. *Assessment of iodine deficiency disorders and monitoring their elimination- A guide for programme managers* (third edition). Geneva, WHO. Available at: http://apps.who.int/iris/bitstream/10665/43781/1/9789241595827_eng.pdf

89 FAO. 2014. *Food and Nutrition in Numbers 2014*. Rome, FAO. Available at: www.fao.org/ publications/card/en/c/9f31999d-be2d-4f20-a645-a849dd84a03e (accessed 29.08.2016)

Notes